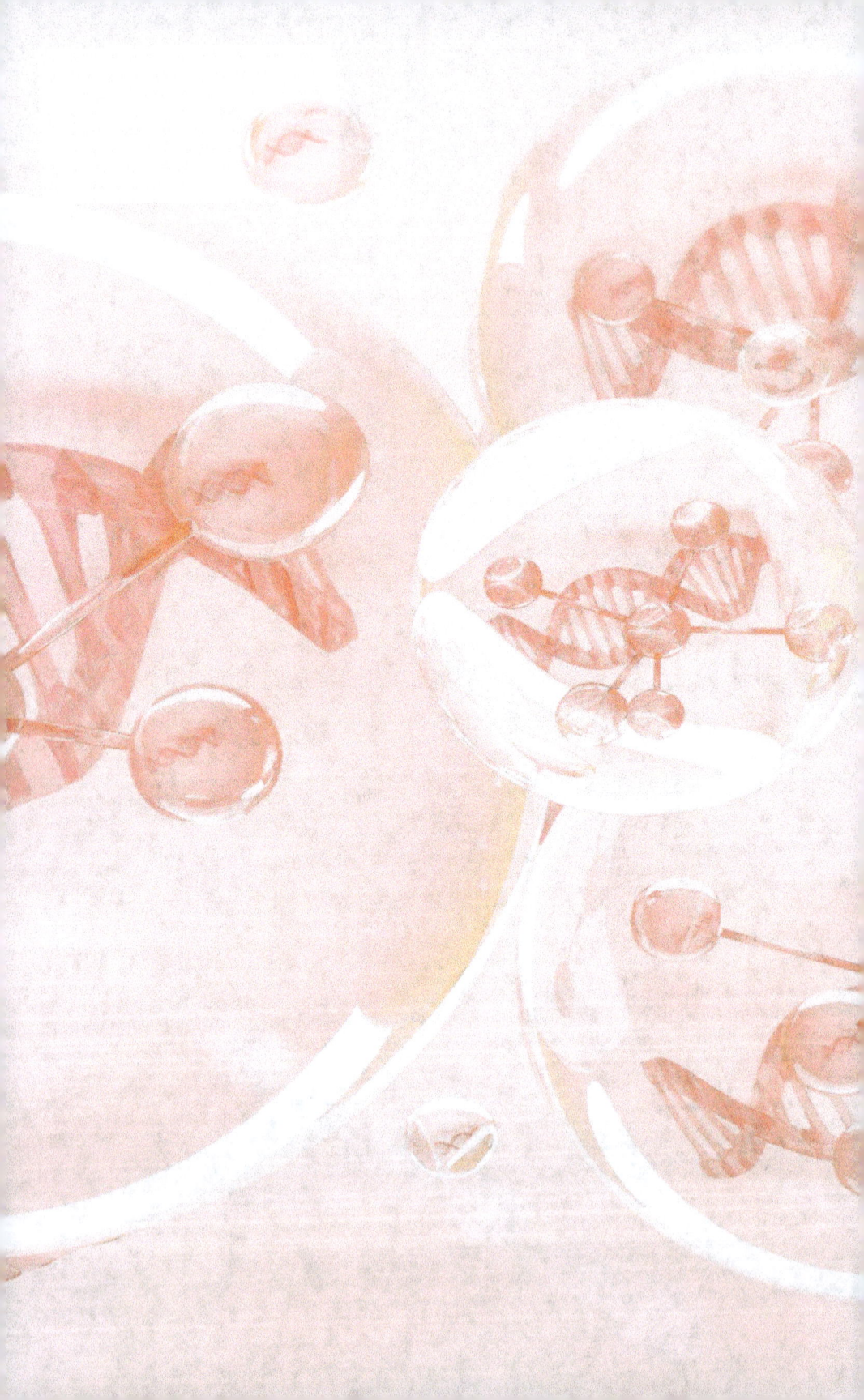

Each one with its trillions of variables

"The human being is nothing more than a set of variables, forged in the cosmos, expressed on earth, assembled in the womb and known in the mind." **Paul McCullough**

I dedicate this book to my children.

Will and Bjørn

Thanks

Érika Lorente Neves, for her work in reviewing and directing the structures of literary content.

Preface

Are we here by chance? The trajectory that brought us to the current human being is a complex narrative, driven by an evolution guided by countless cosmic variables, from the forces that shaped the Earth to the biochemical events that gave rise to life. Throughout the eons, evolution has sculpted living beings, perfecting every detail in response to environmental factors.

Consciousness has emerged as a unique expression of this evolution. This book explores the specific variables that culminated in the emergence of modern humans, highlighting ideal geophysical conditions, genetic complexities, and other elements that contributed to the formation of Homo sapiens.

Intelligence, reflection, and curiosity are intricate outcomes of a convergence of cosmic variables over time. As we explore the pages of this book, we embark on a journey through the complexity of human evolution, examining decisive moments and fundamental interactions that transformed us into beings capable of contemplating our own origins and understanding the forces that shaped us. The narrative highlights the extraordinary evolution that has brought us to the current point, where the complexity of the cosmos is reflected in the human mind.

Michel Houelle

Index:

Introduction

The motive of this book and its audience.

Hard work is not synonymous with, or even the representation of, what is inflexible. Be it learning, work, management, studies, or any arduous activity. On the contrary, arduous is complex and because it is so, it tends to have better dynamics and greater possibilities of flexibility. Real villains are the easy ways. The path devoid of variables has no options but rigidity; That's why we say, "Let's keep it simple." The simple is the most rigid and it is a place of no return. No one wants anything that isn't simple, simple is the norm; "Don't complicate it, leave it in a normal way", that's what I often hear.

My reflection here in this book is to demonstrate that absolutely nothing is simple. The whole simple path is insufficient, devoid of options.

Ignoring the details, levelling down is the strategy for everyone to understand and accept. For this to occur, the complex is deleted. Globalization causes simplification and what is complex is rejected as a villain. Moreover, if the complex path challenges the easy path, it will lose politically. It is wrong to correct the errors of simplicity.

Something like this:

"Yes, it's wrong, but don't say anything, rebutting this idea will only create controversy."

"There are studies in this area, it's enough to know that it's what has worked, and we're going to leave it as it is."

"Thousands adhere to these ideas, which means that the ideas are right".

Putting variables in boxes and classifying the complex as a set is a practice for those who tend to shallow thinking, to lack of understanding of the true fabric of existence, where believing is easier than knowing. Canned joints without reflection can life and no one can think anymore, everything is easier to manipulate, it is easier than innovating and creating.

It is never right that what prevails is the lack of adequate knowledge; To remain silent in the face of ignorance and the tendency to be easy is to be complicit in the consequences that will follow. It is at this point that critical thinking plays a key role in promoting respect for human variability.

Critical thinking provides the individual with the ability to question assumptions and stereotypes, which are prejudices that facilitate generalizations, contributing to a deeper understanding of the diversity of experiences, perspectives, and values.
Critical skills enable people to make fairer decisions, considering factors such as diversity, inclusion, and social justice, when making decisions in all spheres, to talk and defend their views in a respectful and reasoned way.

In addition, it is a way to gain a deeper understanding of the experiences of others, which increases empathy and meaningful connections.

Critical thinking empowers people to question stigmas, stereotypes, and discriminatory practices, which contributes to the creation of fairer and more inclusive environments.
It pushes you to learn every time. It is crucial to understand and respect human variability, which is dynamic and evolves over time, allowing people to adapt their approaches, strategies, and interactions as nuanced and complexities while avoiding rigid and inflexible simplistic approaches.

In my profession there are important and constant questions about the management of psychological problems, health and behavioural problems, adaptive, social relationships, etc. However, it is the other written above that ends up being the bulk of the work. It is the diversity and the great role of variables that give such a good challenge to those who work with people.

What is the audience for this book? Here the statements are subject to constant revision, also all the information is condensed, and each topic has the potential to be a book. The reader should not take everything for granted but use their own tools to continue investigating and expanding their knowledge; Read the references I put here and draw your own conclusions.
I'm aware language isn't always easy, maybe dense, it's exactly what I want, to share with you my knowledge, so I'm not going to make it easy. I challenge you to exercise your great capacity to assimilate.

I want to share my perspective on the complexities of each person and the possibilities of adapting the best interactions to develop growth.

Human variables are the main reason for my text. The public is the one that is scientifically curious and wants to read more about the topic.

Happy reading

Paul McCullough

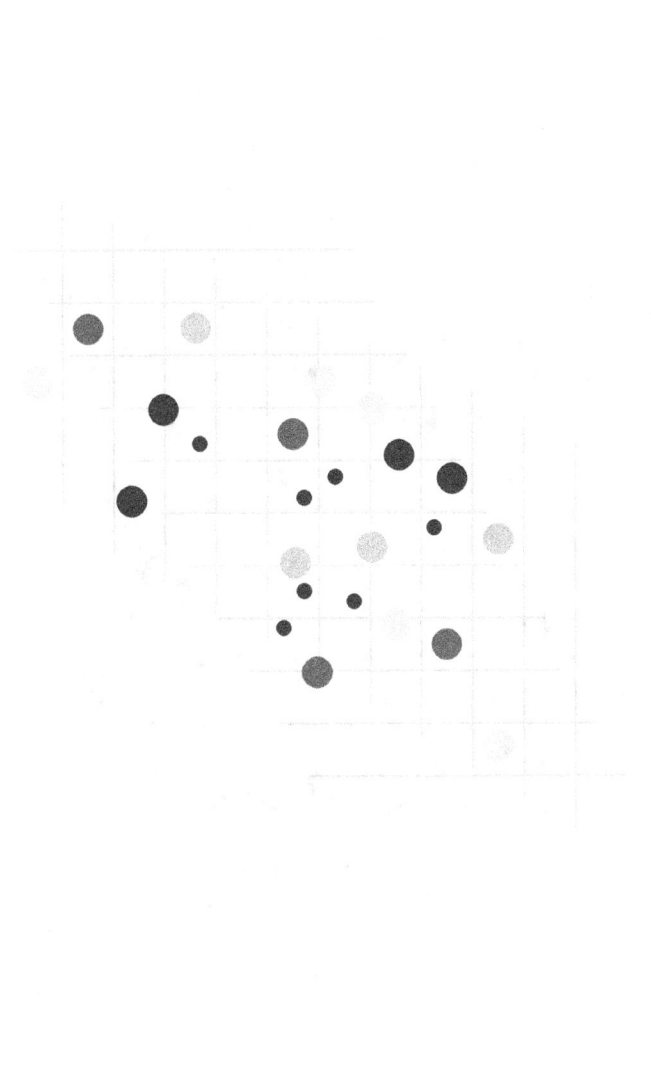

Chapter 1

The variables.

"Nothing is permanent except change."
Heraclitus

Variables, what are they?

Variables are elements that do not have consistency or a fixed pattern, which implies that they can adopt different values or change over time. This characteristic makes it possible for them to represent a wide range of situations or scenarios. In other words, variables could assume different numerical values or states in response to different conditions, inferences, or interferences. A variable, by nature, is not restricted to a single value or state; It is flexible and can be manipulated or unmanipulated as needed.

Variables are the basis of all development. Science would not exist without the understanding of variables, moreover, the human being is both historically and in his current life, made up of countless variables.

When I use the term "development" I refer to the transformation or evolution of something over time, influenced by several variables and factors that shape the outcome. This can apply to anything that is subject to change, progress, or growth in some aspect. It is a broad and fundamental concept in various areas of life, of society, in short of everything that exists.

When I think about the possibility of extraterrestrial life like ours, I soon understand that it would be almost impossible to have variables so like ours that would allow the evolution of humanoid life, or beings minimally like human beings. The same conditions, even the pre-existence of life, would need to be absolutely the same as those by which our planet was formed.

Variables and their controls are the basis of science.

Science observes and uses different types of variables that serve to measure, analyse, and understand natural phenomena and complex systems. At this point, the variables are not just a point of philosophical observation, but also serve as tools for conclusions when observed with control.

Here are some of the types of variables common in different areas of science, I will mention the ones that I know closest to my area of expertise, but certainly each branch of science has a range of variables that are its basis for studying and understanding what it studies.

My favourite variable of all the sciences is the independent variable. It doesn't depend on the variations of the variables, only I can control it, if you don't choose this variable well your research may come to nothing. It's always a challenge and that makes it very interesting to me.

Natural Sciences:

Dependent variables: These are those that are measured or observed in an experiment or study. Examples include temperature, pressure, concentration of substances, velocity, mass, etc.

Independent variables: These are those that the researcher manipulates or controls to see how they affect the dependent variables. For example, the time of exposure to sunlight in a study on plant growth.

Controlled variables: These are variables that are kept constant in an experiment to ensure that only the independent variable affects the results.

Extrinsic variables: These are external factors that can affect the experiment but are not variables of interest. These factors should be controlled or considered.

Intrinsic variables: These are characteristics inherent to the system or object of study that affect the results. For example, the age of an individual in a study on cognitive performance.

Social sciences:

Dependents: These can include attitudes, behaviours, opinions, levels of satisfaction, among others, that are observed or measured in social studies.

Independent: These are often social, demographic, or environmental characteristics that researchers believe influence dependent variables.

Control variables: These are used to control or eliminate the effect of variables that may interfere with the results.

Contextual: These refer to the environment or context in which social studies are conducted, such as geographic location, culture, time, etc.

Health Sciences:

Clinical variables: These include medical parameters such as blood pressure, blood glucose levels, pulse, temperature, among others.

Outcome variables: These can be measures of improved or worsened health, such as recovery from an illness, reduction of symptoms, quality of life, etc.

Exposure variables: These represent risk factors or exposures that may be associated with health conditions such as smoking, diet, physical activity, etc.

Computer Science:

Input variables: These are data or information provided to a program or algorithm.

Output variables: These are the results or responses generated by the program or algorithm based on the input variables.

State variables: These represent internal information that a program or system uses to maintain its functioning, such as control variables, counters, etc.

Evolution and variability are intrinsically linked. Evolution is like building a house with the same plant and materials, but on different terrains, such as an arid desert or a wetland. These conditions influence the durability of the house, the strength of the materials, and their degradation over time.

Similarly, all life on Earth follows a building code, but this construction varies depending on different factors such as time and location. This text focuses on human life and how these variables have shaped our existence throughout history."

For example, if planet Earth were closer to the sun by only 1 meter, the Earth would receive slightly more solar radiation, which would lead to a modest increase in temperature. This change may not be immediately noticeable, but over time, it could contribute to warming trends, preventing certain bacteria from being able to survive, or the faster evaporation of water from the oceans. We will see in a future chapter that our oxygen depended on oceanic, climatological, meteorological aspects, etc. The exact combination of these variables provided the conditions for cyanobacteria and other elements to produce atmospheric oxygen and eukaryotic cells, plants, etc.

The change in distance would affect Earth's elliptical orbit, potentially altering the duration and intensity of the seasons. Summers can get hotter, and winters can be milder, depending on the position of the Earth in its orbit at any given time. This altered orbit could disrupt established weather patterns, potentially leading to changes in the flow of ocean currents, temperatures, wind flow, etc. It would have ripple effects on regional climates and ecosystems. Even small changes in Earth's environment can have far-reaching effects over geologic timescales.

Nowadays we are manipulating a dangerous independent variable.

Let's use an analogy to demonstrate the dangers of mishandling the independent variable, in the context of greenhouse gas emissions and climate change:

Imagine that you are conducting a study on the effects of different diets on human health, with the independent variable being the type of food consumed. If you decide to manipulate this variable inappropriately, such as offering only high-calorie, high-saturated fat fast foods to one group while feeding another group a balanced, healthy diet, the results of your study could be misleading and dangerous.

Similarly, in the context of climate change, greenhouse gas emissions, they can be considered the Earth's "diet." If a researcher manipulates these emissions intentionally, for example by artificially increasing them, the results of the study on the effects of climate change could be affected. That would be like feeding the Earth the wrong way, which could lead to wrong conclusions about the effects of greenhouse gases.

I guess I used a sarcastic analogy, didn't I? We know what we're doing to the earth, don't we?

However, any inappropriate manipulation of such an important variable by those dedicated to minimizing the impacts of climate change can result in underestimating the real effects of emissions, obscuring understanding of the detrimental effects on climate and biodiversity, and potentially leading to inappropriate choices about policies and methods to curb climate change. In the same way that an inadequate diet can have a negative impact on human health, an improper handling of climate variables can harm the health of the planet and compromise our ability to deal with climate change effectively.

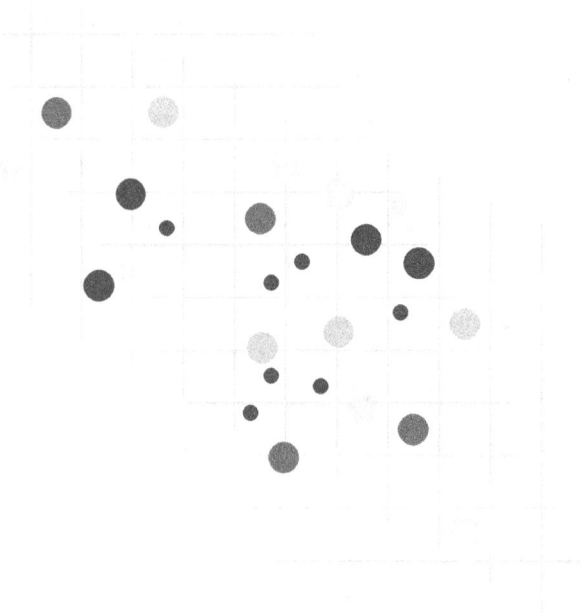

References chapter 1:

1. Creswell, J. W. (2014).
2. Research Design: Qualitative, Quantitative, and Mixed Methods Approaches. Sage Publications.
3. Yin, R. K. (2018).
4. Case Study Research and Applications: Design and Methods. Sage Publications.
5. Kirkman, T. W. (1996).
6. Statistics to Use. Academic Press.
7. Anderson, D. R., Sweeney, D. J., & Williams, T. A. (2019).
8. Statistics Applied to Administration and Economics. Cengage Learning.
9. Neuman, W. L. (2014).
10. Social Research Methods: Qualitative and Quantitative Approaches. Pearson.
11. Bryman, A. (2016).
12. Social Research Methods. Oxford University Press.
13. Portney, L. G., & Watkins, M. P. (2015).
14. Foundations of Clinical Research: Applications to Practice. F.A. Davis Company.
15. Polit, D. F., & Beck, C. T. (2017).
16. Nursing Research: Generating and Assessing Evidence for Nursing Practice. Wolters Kluwer.
17. Han, J., Kamber, M., & Pei, J. (2011).
18. Data Mining: Concepts and Techniques. Morgan Kaufmann.
19. Bishop, C. M. (2006).
20. Pattern Recognition and Machine Learning. Springer.

Chapter 2

In the beginning, the variables varied.

"The cosmos is a vast, complex, and interconnected fabric of events, and our understanding of the universe is just beginning." **Carl Sagan**

In the beginning

Human life is the result of a complex interplay of numerous variables and factors, and it is challenging almost impossible to pinpoint an exact number of these interactions. The conditions and variables that have made human life possible are diverse and span multiple scientific disciplines.

The position of our planet in the solar system is what makes possible the conditions for life as we know it, this locality is often referred to as the "habitable zone" or the "Goldilocks zone". Do you remember the story? The girl who found the ideal "porridge", which was neither too cold nor too hot, ideal to be able to eat, in the same way we are exactly where life can occur.

The temperature on Earth is within a range that allows water to exist in its liquid form, which is essential for life as we know it. If a planet is too close to its star, it becomes too hot for liquid water, it evaporates, and if it is too far away, it becomes too cold, causing the water to freeze.

The Earth's atmosphere provides the necessary mixture of gases, mainly oxygen and nitrogen, that living things need to breathe. Also, water serves as a solvent for biochemical reactions, which is another temperature-regulating factor providing a habitat for many organisms. Earth's gravitational pull is essential for holding our atmosphere in place and maintaining stable conditions on the planet.

The Sun provides the energy needed for photosynthesis, which forms the basis of most terrestrial food webs. It also provides warmth and light. A stable, long-lived parent star (in our case, the Sun) is necessary for the evolution of complex life.

The Magnetosphere, or Earth's magnetic field, protects the planet from harmful solar radiation and helps maintain a stable climate.

Carbon-based chemistry is the building block of organic molecules, which are the building blocks of life. Nutrient cycles, such as the carbon, nitrogen, and water cycles, are essential for maintaining ecosystems and providing the resources necessary for life.

Plate tectonics play a role in regulating climate and maintaining the planet's geological activity.

Oxygen crucial for aerobic respiration in many organisms, including humans, and a relatively stable climate with moderate temperature variations is essential for long-term habitability.

Later, after the appearance of what we consider living organisms, we can talk about the consequences that life itself provided by being on a planet in the "habitable zone", such as the genetic diversity that within species allows adaptation and evolution over time. Life depends on complex biomolecules like DNA, RNA, proteins, and lipids for cellular processes.

Biodiversity, which is a diverse range of species and ecosystems, provides resilience and stability to the biosphere. Evolutionary processes, which are the evolutionary mechanisms, drive the development and adaptation of life forms over time.

Life on earth and the co-evolutionary interaction between life and the planet's environment.

Today, we live in a world where both the atmosphere and oceans are rich in oxygen, supporting a great diversity of life, including plants, animals, protozoa, algae, and bacteria. This is very different from the early days of Earth when there were only simple bacteria-like organisms and no oxygen in the atmosphere.

We know that the Earth had no oxygen about three billion years ago. We don't have direct samples of the atmosphere of that time. However, iron-rich sedimentary rocks, known as banded iron formations, provide some clues. These rocks cannot form in today's oxygen-rich oceans, as the iron in them needs to be transported without reacting with oxygen (yes, iron would rust, or oxidize). This suggests that the early oceans were poor in oxygen, which also affected the biology of the time. Other evidence, such as the presence of iron pyrite in ancient sedimentary deposits and the chemistry of rocks exposed to the atmosphere, also point to an oxygen-free environment.

The water in the oceans was very poor in oxygen and remained in liquid or gaseous form. However, always water. Oxygen remained in small quantities attached to water, and in the composition of some rocks, and not in the form of gas as exists today. Humans could not live on earth at those times, The transition to an oxygen-rich atmosphere occurred gradually about 2.4 billion years ago, allowing for the emergence of more complex life forms that rely on oxygen.

The first organisms on early Earth were probably anaerobic, meaning they did not rely on oxygen for their metabolic processes. They could utilize alternative energy sources, such as chemical reactions that did not involve oxygen. These simple organisms were able to carry out the basic processes of life, such as DNA replication and protein production, without the harmful (oxidizing) interference of oxygen.

As oxygen began to be produced in significant quantities, mainly by photosynthesis by cyanobacteria, it was gradually released into the atmosphere and oceans. This slow and gradual addition of oxygen allowed for the development of more complex life forms, such as eukaryotic cells. Eukaryotic cells have a distinctive feature: a membrane-separated nucleus that houses genetic material. This enabled more precise regulation of cellular processes and greater complexity.

In addition, oxygen played an important role in the oxidation of chemical compounds and reduced gases, such as hydrogen and hydrogen sulphide, produced by certain bacteria. This process released energy and paved the way for a diversity of life forms, including multicellular organisms.

Therefore, the initial absence of oxygen created an environment conducive to the emergence of life and the first organisms, while the gradual addition of oxygen into the atmosphere and oceans allowed for the development of more complex and diverse life forms.

Earth's history shows that the evolution of life and the environment are intertwined, with changes in the environment driving the evolution of life and vice versa.

The co-evolution between life and the environment is a key feature of Earth's history, and understanding this gives us a new perspective on our own role on the planet, where complex life is heavily influenced by the bacteria that coexist with us.

How many variables have you counted so far? cosmological, chemical, chronological, biological, geological, thermic, meteorological, magnetic, energetic, etc. All of them, acting in unison to bring to Earth the conditions of its development.

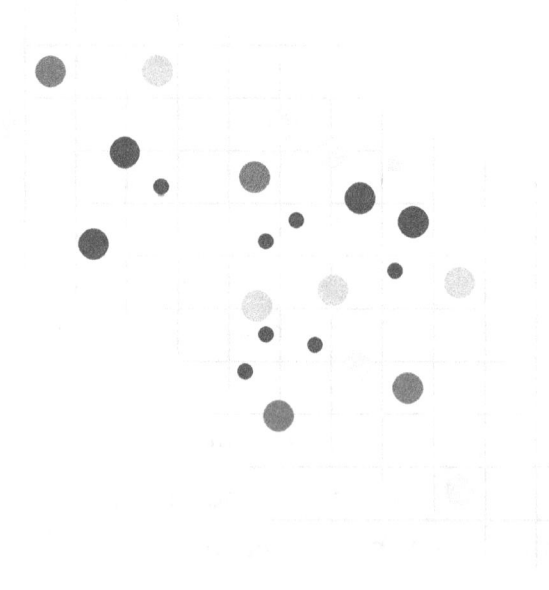

References chapter 2:

1. *Kasting, J. F., et al. (1993).*
2. *Habitability of Planets Around Red Dwarf Stars. Icarus, 101(1), 108-128.*
3. *Raven, P. H., et al. (2007).*
4. *Biology: Eighth Edition. McGraw-Hill Education.*
5. *Sagan, C. (1974).*
6. *The Cosmic Connection: An Extraterrestrial Perspective. Doubleday.*
7. *Russell, C. T., et al. (2016).*
8. *The Magnetospheres of Earth and Jupiter: Comparisons and Contrasts. Journal of Geophysical Research: Space Physics, 121(11), 11,057-11,071.*
9. *Smith, J. M., et al. (2008).*
10. *Principles of General, Organic, and Biological Chemistry. McGraw-Hill Higher Education.*
11. *Schlesinger, W. H. (1997).*
12. *Biogeochemistry: An Analysis of Global Change. Academic Press.*
13. *Turcotte, D. L., & Schubert, G. (2002).*
14. *Geodynamics. Cambridge University Press.*
15. *Widmaier, E. P., et al. (2016).*
16. *Vander's Human Physiology: The Mechanisms of Body Function. McGraw-Hill Education.*
17. *Knoll, A. H. (2014).*
18. *Life on a Young Planet: The First Three Billion Years of Evolution on Earth. Princeton University Press.*

Chapter 3

Human development, the Origami of life.

"In the womb, the soul prepares for its earthly journey, absorbing knowledge of the cosmos before emerging into the light of the world." **Plato**

Emerging

While it is tempting to explore the development of humanity from its beginnings millions of years ago to the present day, I will not address that question at this time; Maybe we can do it another time. So, I'm not going to go into hominids or the evolution that led us to Homo sapiens etc.

However, I intend to address the differences and variables to which the human body, as we know it today, is subject. I will focus my attention primarily on the human brain and the formation of everyone, as well as its remarkable variables. These differences are widely studied in fields such as psychology and neuroscience. Therefore, this is the central theme of this brief text.

What makes us human is our brain and from it, the evolution of each person.

How do we get started?

First, I would like to point out that the following text is an explanation of how we start and the variables to which we are subject.

Before exploring the more scientific topic, we should address some current theories:

Theory number 1 - It's the storks that bring the babies. There are numerous variables that can influence the appearance of a baby, such as the wind at the time of the stork's take-off, the outside temperature, and even the zip code where it will deliver the baby. For example, streets without a zip code are a problem, also if the code is wrong, as well as the issue of air traffic and stork strikes.

Theory number 2 – Daddy plants a little seed in Mommy's belly, but I have some resistance to exploring this topic. Variables can include the amount of soil in mommy's belly and even the amount of water we should water mommy's feet during the gestation period, not to mention insecticides against aphids.

Leaving aside the various theories above, let's focus on what we know.

The variables of each person's construction depend on genetic and environmental factors, nutrients, psychological factors, biochemical factors of all kinds. These factors can affect embryonic development at any time during pregnancy.

Genetics play a key role in the development and functioning of the human body. Genes are segments of DNA that contain information encoded to make proteins and perform specific functions in the body. These proteins play critical roles in various bodily functions, from embryonic development to daily functioning.

During embryonic development, genes are turned on and off at specific times to control the formation of important structures such as the notochord, neural plate, and many others. The notochord is a temporary structure that plays a key role in inducing neural tube formation, which subsequently develops the central nervous system. Specific genes are responsible for this sequence of events.

Furthermore, throughout life, genes continue to play essential roles in our health and well-being, regulating processes such as growth, tissue regeneration, disease response, and the maintenance of vital functions in our body. Therefore, we can say that genes are like the instructions encoded in our DNA that control virtually every aspect of our biological development and functioning.

Embryonic development involves several important stages that can be related to approximate weeks of gestation, although it is important to note that these estimates may vary slightly from one baby to another.

Fertilization (0 weeks): This happens when a "sperm" meets an "egg" to form a "zygote." If you still don't know how fertilization occurred, you can stick with one of the theories listed at the beginning of this text.

Segmentation and morula (1-4 weeks): The zygote divides several times to create a group of cells called a "morula."

Blastocyst (5-6 weeks): The morula turns into something called a "blastocyst," which has two parts: one that will become the "placenta" and one that will be the "embryo."

Implantation (6-7 weeks): The blastocyst attaches to the wall of the uterus and begins to create the placenta and future baby.

Gastrulation (2-3 weeks after implantation, about 8-9 weeks after conception): Here, different parts of the body begin to form from the cells of the embryo.

Neurulation (3-4 weeks after implantation, about 9-10 weeks after conception): At this stage, the central nervous system begins to develop from the cells of the embryo.

Organogenesis (after neurulation, over the course of several weeks): Now, the body's organs and systems begin to form from the germ layers.

Ongoing foetal development (until birth, usually around 40 weeks after conception): During this time, the baby grows and becomes more complex, developing parts such as arms, legs, hair, and nails.

Complete closure of the foetus: It is not a specific stage, but it can refer to times when parts of the body close, such as the development of the nervous system or internal organs.

Keep in mind that these are rough estimates, and development may vary between people. The total development time is about 40 weeks from conception, but we usually calculate the age of pregnancy from the last menstruation, which adds about two weeks to the calculation.

Here the interesting thing is the steps that are subjected to genetic programming:

Neural plaque formation: Neurulation begins with the induction of the neural plaque, which is a flat, thickened region of the ectoderm that forms along the dorsal midline of the embryo. This initial step is influenced by signalling molecules from adjacent tissues, such as the notochord and the underlying mesoderm.

Neural sulcus formation: As the neural plate continues to develop, it undergoes an invagination process. The edges of the neural plate rise and bend inward, forming a groove along the midline known as the neural sulcus.

Neural fold elevation: The neural folds on either side of the neural sulcus elevate further as development progresses. These neural folds are created because of differential growth in the dorsal ectoderm.

Neural Tube Closure: The most critical step of neurulation is the closure of the neural tube. The neural folds continue to move towards each other until they eventually meet and fuse into the dorsal midline. This fusion process proceeds in the cranial (towards the head) and caudal (towards the tail) directions. When the neural tube is fully closed, it becomes a hollow, tube-like structure that is separated from the rest of the embryo.

Neural Crest Cell Formation: During neural tube closure, a group of cells at the dorsal edge of the neural tube do not undergo complete fusion. These cells are known as neural crest cells. They migrate to various parts of the embryo and give rise to a diverse range of cell types, including sensory neurons, autonomic neurons, and various types of connective tissues.

Proper neurulation is essential for the development of a functioning central nervous system. Neurulation defects can lead to serious congenital disorders, such as neural tube defects (NTDs), which include conditions such as spina bifida and anencephaly. These disorders result from failures in neural tube closure during embryonic development. Prenatal care, including folic acid intake, is important in preventing NTDs and ensuring proper neural tube development.

The genetics involved in neurulation and neural tube formation are complex and influenced by multiple genes and signalling pathways. Here are some key genetic factors and signalling pathways that play essential roles in this process:

Sonic Hedgehog (Shh) Signalling: Sonic Hedgehog is a critical signalling pathway that plays a central role in neural tube development. It is secreted by the cells of the notochord and the floor plate and is essential for patterning the neural tube along its dorsoventral axis. Shh signalling helps establish different regions of the neural tube and controls the differentiation of neurons and other cell types within it.

Bone Morphogenetic Protein (PMO) Signalling: PMOs are a family of growth factors that are involved in regulating neural plate and tube development. PMO signalling promotes the formation of the non-neural ectoderm and inhibits the formation of neural tissue. A balance between PMO and Shh signalling is critical for proper neurulation.

Nodal Signalling: Nodal is another signalling pathway that is involved in establishing the left-right axis of the embryo. Proper left-right patterning is crucial for the correct positioning and orientation of the neural tube.

Transcription factors: Several transcription factors are key pieces in neural tube development. For example, the Pax family of transcription factors is involved in specifying different regions of the neural tube. Pax3 and Pax7, for example, are important for the development of the dorsal neural tube.

Genetic mutations: Genetic mutations in specific genes can lead to neural tube defects (NTDs), which are congenital conditions in which the neural tube fails to close properly during development.

Folate metabolism genes, such as MTHFR, are of particular interest in the context of NTDs as they can affect the availability of folate, a nutrient crucial for neural tube closure.

Genomic variants: In addition to specific genes, variations in the genome, including peripheral nervous system single nucleotide polymorphisms (SNPs), may influence the risk of NTDs. Some of these variants may affect the function of genes involved in neural tube development or folate metabolism.

Environmental factors: It is important to note that genetics alone do not determine the risk of NTDs. Environmental factors, such as maternal diet and exposure to certain substances, can interact with genetic factors to influence the risk of NTDs.

The genetics of neurulation is a complex interplay of multiple genetic pathways and factors, and research in this field continues to uncover new insights into the molecular mechanisms underlying neural tube development and associated congenital disorders. Understanding these genetic aspects is crucial for preventing and addressing neural tube defects and other central nervous system abnormalities.

Foetal cognition:

Still here, I would like to add a theme that I have often found on social media for the public, articulators of pseudoscience who are dedicated to affirming and basing emotional consciousness on embryos and foetuses throughout the intrauterine period.

The development of the nervous system in the embryo is a complex and fascinating process, the nuances of which have been thoroughly explored in several scientific sources. The book

"Principles of Developmental Biology" by Scott F. Gilbert offers a comprehensive overview of this phenomenon, while the article "Human brain development: cell by cell mapping of the early human foetal brain" (Nature Neuroscience, 2019) deepens the understanding by mapping the human foetal brain in early stages cell by cell.

The formation of the cerebral cortex, as discussed in the article "Development of the human cerebral cortex: Boulder Committee revisited" (Nature Reviews Neuroscience, 2001), is a crucial aspect of this development, requiring time and precision for its configuration. This process involves not only cell differentiation but also the migration of cells to the proper positions, an intricate cellular ballet essential to brain architecture.

Regarding cell migration and neural development, the article "Neuronal migration and its disorders" (Molecular Psychiatry, 2007) offers valuable insights into this complex phenomenon, highlighting the importance of the correct positioning of cells for the proper formation of neural structures.

In postnatal development, as discussed in the article "Development of the prefrontal cortex: Mechanisms and implications for cognitive development" (Neuroscience & Biobehavioural Reviews, 2004), the preponderance of the prefrontal region emerges in cognitive development. This stage is intrinsically linked to cognition and emotion, highlighting the interconnectedness between structural development and the emergence of higher cognitive functions.

It is imperative to note that although the human brain begins to form in the early stages of embryonic development, the structures responsible for advanced cognitive functions take longer to develop. It is important to emphasize that the foetus, before birth, does not have the fully developed systems to make sense of or understand cognitively all the variables that give rise to cognitive thinking. In addition, embryos lack the ability to feel emotions, since emotions require cognitive sense and input from external stimuli. This limitation is due to the incompleteness of the receptor organs and the absence of neuroplasticity during the embryonic stage.

The establishment of synapses and neural networks, as discussed in the article "Synapse Formation in Developing Neural Circuits" (Current Opinion in Neurobiology, 2001), is a crucial stage for brain functionality. The formation of these connections is essential for the efficient transmission of signals between nerve cells, laying the foundation for complex cognitive and emotional functions.

Certainly, the assertion that cognitive and emotional capacities only fully manifest after birth is based on an understanding of the development of the central nervous system. This process is complex and involves several phases, from birth to the consolidation of brain connections during the first years of life.

During embryonic development, neural cells begin to differentiate and form the nervous system. As the foetus develops, nerve cells multiply and migrate to their appropriate positions. This is followed by synapses, or connections between nerve cells, which are critical for the transmission of electrical information and communication between different areas of the brain.

However, it is crucial to note that although the structure of the nervous system is present before birth, cognitive and emotional functions only fully develop after childbirth. The postnatal period plays a crucial role in this process, since exposure to external stimuli, social interactions, and experiences affect brain development and affect cognitive and emotional skills.

Criticism of scientific constructions

As we have seen, there is a known process for the brain to receive and process information after birth. To seek theories to adapt to a belief is to encourage some to continue looking for a way to include cognitive emotions in foetal development, which would be manipulation of data. I want to make it clear, those who use manipulated studies, or who are not responsible for the conclusions of the studies mentioned, cannot be included as the author of this manipulation, users in their vast majority are passive and exempt from these results.

Interpretations that are not clear should be criticized. The emphasis on the relevance of the scientific criterion is imperative. Science will understand the world objectively and based on evidence. Claims that are not scientifically evaluated.

Evidence is data, and manipulating data to gain reason is anything but science. False science is often based on incomplete information, saying that there are ongoing studies or that this or that is an area not yet defined by science is an excuse to continue using incomplete methods, without selective interpretations of data or absence of consistent empirical evidence.

To ensure that theories and claims about cognitive and emotional development are scientifically grounded, it is imperative that there be a rigorous review by the scientific community, to avoid the dissemination of unfounded or potentially harmful information.

History demonstrates that being a scientist, launching and completing scientific studies, still working on something, but all of this is not synonymous with validation. Below I will cite examples of recognized scientific papers that have had a negative influence, either due to lack of validation or even due to unscrupulous manipulation of their variables.

Andrew Wakefield and the Supposed Link Between the MMR Vaccine and Autism (1998): Andrew Wakefield's study, published in The Lancet, suggested a link between the MMR vaccine (measles, mumps, and rubella) and autism. The study was later discredited, and the falsification of data was discovered. This incident had significant repercussions, including decreased vaccination rates and outbreaks of vaccine-preventable diseases.

Here is a case of devastating consequences for the use of a drug, thalidomide. The drug was initially marketed as a safe drug to relieve nausea and induce sleep, especially in pregnant women.

The thalidomide scandal occurred mainly during the 1950s and early 1960s. The drug has been linked to a few serious birth defects in newborns whose mothers had taken it during pregnancy. Many of these children were born with malformations in the upper and lower limbs, known as phocomelia.

Nowadays, every time a pseudoscientific process is found, the scientific community tries to create control mechanisms so that such irresponsibility does not occur again. However, ideas of all kinds continue to circulate that still need validation.

Establishing something as science is a huge responsibility, it affects life. It cannot be done for illegitimate purposes.

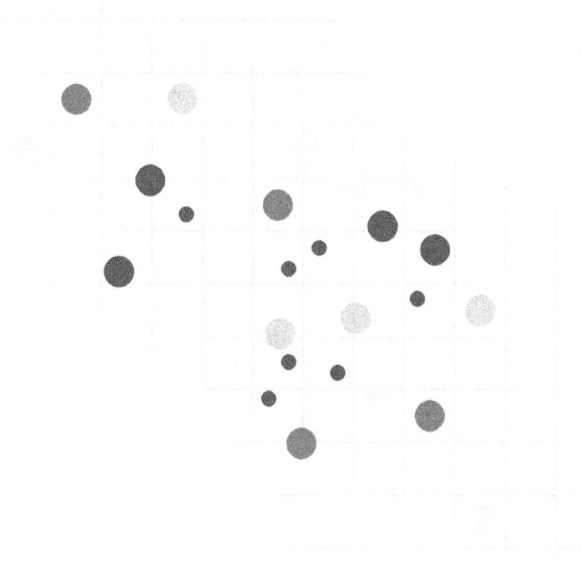

References chapter 3:

1. Gilbert, S. F. (2014).
2. Developmental Biology. Sinauer Associates.
3. Wolpert, L., Beddington, R., Brockes, J., Jessell, T., Lawrence, P., & Meyerowitz, E. (2015).
4. Principles of Development. Oxford University Press.
5. Harris, M. J. (2019).
6. "Neural Tube Defects." In Handbook of Clinical Neurology (Vol. 162, pp. 255-275). Elsevier.
7. Greene, N. D., & Copp, A. J. (2014).
8. "Neural tube defects." Annual Review of Neuroscience, 37, 221-242.
9. Copp, A. J., & Greene, N. D. (2013).
10. "Genetics and development of neural tube defects." Journal of Pathology, 220(2), 217-230.
11. Detrait, E. R., George, T. M., Etchevers, H. C., & Gilbert, J. R. (2005).
12. "Human neural tube defects: developmental biology, epidemiology, and genetics." Neurotoxicology and Teratology, 27(3), 515-524.
13. Dessaud, E., & Briscoe, J. (2009).
14. "Sonic hedgehog signaling in vertebrate neural tube development." Cell and Tissue Research, 337(1), 49-66.
15. Bery, A., Martynoga, B., Guillemot, F., & Joly, J. S. (2010).
16. "Pax3/Pax7 mark a novel population of primitive myogenic cells during development." Genes & Development, 24(13), 1426-1431.
17. Shaw, G. M., & Finnell, R. H. (2010).
18. "Blaming genes for human malformations: on the scientific and legal implications of thresholds." Journal of Law, Medicine & Ethics, 38(2), 272-285.
19. MRC Vitamin Study Research Group. (1991).
20. "Prevention of neural tube defects: results of the Medical Research Council Vitamin Study." The Lancet, 338(8760), 131-137.
21. Wakefield, A. J., Murch, S. H., Anthony, A., Linnell, J., Casson, D. M., Malik, M., ... & O'Leary, J. J. (1998).
22. "Ileal-lymphoid-nodular hyperplasia, non-specific colitis, and pervasive developmental disorder in children." The Lancet, 351(9103), 637-641.
23. Deer, B. (2011).
24. "How the case against the MMR vaccine was fixed." BMJ, 342, c5347.
25. Godlee, F., Smith, J., & Marcovitch, H. (2011).
26. "Wakefield's article linking MMR vaccine and autism was fraudulent." BMJ, 342, c7452.
27. Widukind, L., & Widukind, M. (1962).
28. "Thalidomide embryopathy." Archives of Disease in Childhood, 37(194), 505-507.
29. Lenz, W. (1962).
30. "A short history of thalidomide embryopathy." Teratology, 5(3), 79-82.
31. Miller, J. R., Piper, J. M., & Mitchell, J. L. (1973).
32. "Taming the teratogen: A historical perspective on the discovery of the teratogenic effects of thalidomide." Journal of the American Academy of Dermatology, 3(5), 317-323.

33. *Ioannidis, J. P. (2005).*

34. *"Why most published research findings are false." PLoS Medicine, 2(8), e124.*

35. *Nosek, B. A., Spies, J. R., & Motyl, M. (2012).*

36. *"Scientific Utopia: II. Restructuring incentives and practices to promote truth over publishability." Perspectives on Psychological Science, 7(6), 615-631.*

37. *Munafò, M. R., Nosek, B. A., Bishop, D. V., Button, K. S., Chambers, C. D., Percie du Sert, N., ... & Ioannidis, J. P. (2017).*

38. *"A manifesto for reproducible science." Nature Human Behaviour, 1(1), 0021.*

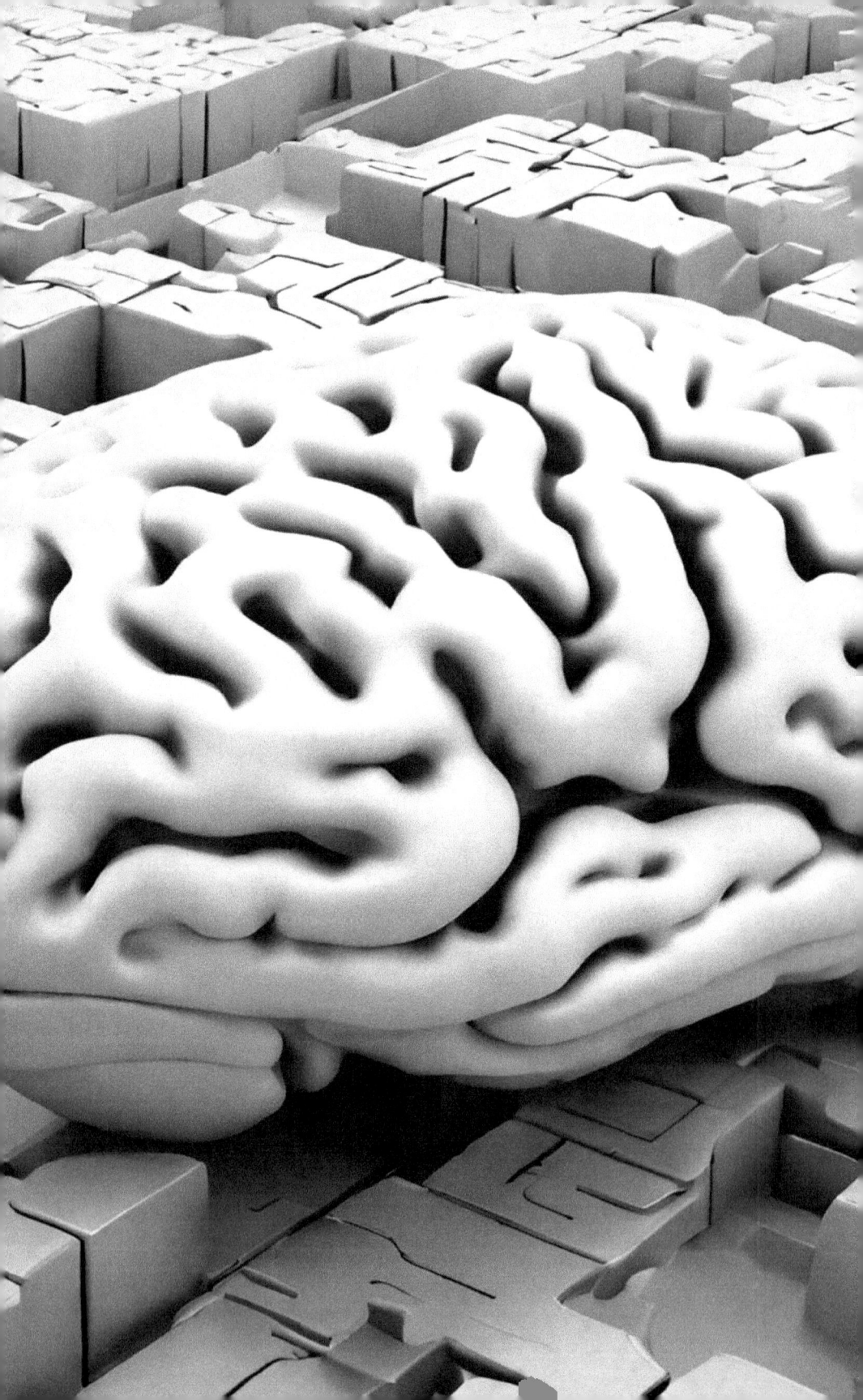

Chapter 4

The variables contained in the formation of the human brain.

"Cogito, ergo sum." **René Descartes**

Building

Brain size is one of the things that has always caught the attention of scientists and researchers in the field of neuroscience and anthropology. It has been a topic of study because of what it might mean for our ability to think and for evolution. Let's take a closer look at the role of brain size in brain development and what it might mean in a simpler way:

It is said that brain size is related to thinking skills. The general idea is that a larger brain offers more resources to nerve cells, and because of this, it can help improve our thinking abilities. But it's important to remember that brain size alone doesn't determine intelligence or how we think. The structure of the brain, the organization, and the number of connections between nerve cells are equally, or perhaps even more, important.

Over the course of human evolution, the size of the human brain has greatly increased compared to our primate ancestors. This increase is linked to the development of advanced thinking skills, such as problem-solving, using language, and complex social interactions. It is considered one of the distinguishing features of human evolution.

The size of the brain in relation to the size of the body, called allometry, is an important thing to consider. It's not just the sheer size of the brain that matters, but also how it compares to the size of the body. This relationship can vary greatly between species and can be an indicator of an animal's niche and cognitive demands.

The concept of neuroplasticity highlights the brain's ability to adapt and reorganize itself. As far as the brain is concerned, "size is not a document" Neuroplasticity is the brain's ability to form new connections and pathways in response to learning and experience, it is what determines the excellence of a brain's functioning. Imagine a very large city, huge buildings, and poor transportation infrastructure. Without proper avenues and streets, we cannot communicate with important areas of the city quickly. Regardless of size, communication and mobility and transport conditions are crucial for the development of the city.

This is the reason why there is no direct correlation between brain size and intelligence. Factors such as genetics, environment, and upbringing play crucial roles in a person's cognitive development. In addition, brain size can vary between different human groups due to genetic and environmental factors.

Advanced brain imaging techniques, such as magnetic resonance imaging (MRI) and functional magnetic resonance imaging (fMRI), have allowed researchers to explore the relationship more fully between brain structure, function, and size. These technologies have helped us better understand how different parts of the brain contribute to specific cognitive functions.

Brain size is a variable with many internal variables.

It is also important to consider evolutionary strategies because they encompass several factors that have shaped the development of the human brain over millions of years of evolution. Understanding these strategies helps us understand how the brain has evolved to meet the challenges and demands of the environment.

Evolutionary strategies have led to the development of specialized cognitive adaptations that confer advantages in terms of survival and reproduction. For example, the evolution of a sophisticated language system has facilitated communication and cooperation within human groups, leading to greater social cohesion and improved collective problem-solving.

Humans are highly social beings, and our evolutionary history has favoured the development of social intelligence. This includes the ability to understand and navigate complex social relationships, recognize facial expressions, and interpret nonverbal cues. Brain regions associated with social cognition have undergone significant development as a result. That is, talking about the development of tools and technology has been a fundamental aspect of human evolution. The brain has adapted to support fine motor skills, hand-eye coordination, and the ability to plan and create tools. These adaptations have led to our ability to use tools complexly, which in turn has influenced our survival and resource acquisition strategies.

Evolutionary strategies favoured cooperative and altruistic behaviours within social groups. The brain has evolved mechanisms that promote trust, empathy, and cooperation, which are essential for successful group living and collaboration.

Humans possess the ability to make long-term plans and decisions that consider future consequences. This cognitive ability, often associated with the prefrontal cortex, has been advantageous in terms of resource management, including food storage, shelter construction, and farming practices.

Another evolutionary variability is the fact that humans are highly adaptable and able to learn from experience. This plasticity allows individuals and populations to adapt to ever-changing environments and challenges over time. The capacity for cultural learning, where knowledge and skills are passed down from generation to generation, has been critical to human survival and development.

As humans have evolved, social structures have become increasingly complex, with larger and more interconnected groups. This complexity necessitated the development of advanced cognitive skills to navigate hierarchies, alliances, and social norms.

These strategies favoured traits and adaptations that improve our capacity for complex thinking, social interaction, tool use, and adaptability. Understanding these factors provides valuable insights into the unique characteristics of the human brain and its place in the natural world. In the context of human brain development, ethological factors can refer to the influence of evolutionary and behavioural patterns observed in both humans and other animals.

Ethology considers how behaviours that have evolved over time in response to specific ecological and social pressures can influence human brain development. For example, humans have evolved social and cooperative behaviours that have likely shaped the development of brain regions associated with social cognition and communication.

Ethology also explores innate or instinctual behaviours that are present from birth. These behaviours may have a neurological basis and are key to understanding early brain development. For example, the seeker reflex in infants is an innate behaviour related to feeding.

Ethological factors consider how the environment, including the social and physical environment, can impact behaviour and, consequently, brain development. For example, studies on child development often draw on ethological principles to understand how early interactions with caregivers influence neural pathways related to attachment and emotional regulation.

Ethology often relies on observational studies to understand behaviour in natural environments. Observations of human behaviour, especially in children, can provide insights into how certain behaviours develop and relate to neural development. And it is this part of science that teaches us that the study of human brain development spans a wide range of disciplines, including neuroscience, psychology, genetics, and others, all working together to provide a comprehensive understanding of the factors that shape the human brain from childhood to adulthood.

Now, let's also insert the physiological factors, here we find the essential components to understand the development and function of the human brain. These factors encompass a number of biological processes and mechanisms that influence the growth, structure, and function of the brain. Here are some key physiological factors related to brain development:

Genetic factors play a key role in brain development. The genetic code in an individual's DNA provides instructions for brain development, including the formation of neural structures, the production of neurotransmitters, and the establishment of neural connectivity. Genetic mutations or variations can have significant impacts on brain development and function, contributing to conditions such as autism, schizophrenia, and intellectual disabilities.

Another variable is neurotransmitters, the chemical messengers that facilitate communication between neurons in the brain. The balance and availability of neurotransmitters, such as dopamine, serotonin, and acetylcholine, are crucial for various cognitive and emotional functions. Imbalances in neurotransmitter levels can lead to mood disorders, such as depression, and affect cognitive processes.

Hormones should also be included in this vast list of variables, hormones are signalling molecules that regulate various physiological processes, including brain development and function. Hormones such as cortisol, released during stress, can have both short- and long-term effects on brain structure and function. Sex hormones, such as oestrogen and testosterone, influence the brain's sexual differentiation during development and continue to affect brain function throughout life.

Neuroplasticity, a very important factor. It is the brain's ability to adapt and reorganize itself in response to learning, experience, and injury. The physiological processes underlying neuroplasticity include synaptic plasticity (strengthening or weakening of connections between neurons) and structural changes in brain architecture. This adaptability is crucial for learning and memory.

The variable of neurotrophic factors and neogenesis.

Neurotrophins are proteins that promote the survival, development, and functionality of neurons. They play a crucial role in brain plasticity, including processes such as the formation of new neurons (neurogenesis).

 Two of the major neurotrophins associated with neogenesis are neural growth factor (NGF) and brain-derived factor (BDNF).

Neural Growth Factor (NGF) is a neurotrophins that plays a crucial role in the development and survival of neurons. It is involved in cell differentiation and the maintenance of neuronal health.

BDNF is a neurotrophins that influences the survival and growth of neurons, as well as playing a crucial role in synaptic plasticity. Synaptic plasticity refers to the brain's ability to adapt to and modify the connections between neurons.

Both NGF and BDNF are related to the stimulation of neurogenesis, especially in the hippocampus, where the formation of new neurons occurs during adulthood. BDNF has been implicated in learning and memory processes.

Therefore, proper regulation of neurotrophic factors is crucial for understanding and promoting neurogenesis, and strategies that stimulate these factors may be beneficial for brain health throughout life and aging.

Neurogenesis occurs during all stages of life; in childhood this variation depends on how stimulating and conducive the environment is for learning. Always including physical activities, which are essential to accompany this process in childhood.

In the case of adults and when healthy, aging not only promotes the accumulation of experiences, memories and knowledge, and cognitive engagement, but also contributes to the promotion of neurotrophic factors, thus favouring neurogenesis, even at very advanced ages.

Keeping the mind active throughout life, through continuous learning, intellectual challenges, and engaging in stimulating activities, provides positive effects on the brain. The constant search for knowledge and adaptation to new situations stimulate brain plasticity and influence the production of neurotrophic factors. However, it is the practice in conjunction with regular physical exercise that promotes these significant benefits for the brain. In addition to improving cardiovascular health, exercise has been linked to increased BDNF levels and improvements in cognition and brain functions.

An enriched environment, which includes social interactions, cognitive challenges, and variety of stimuli, is beneficial. Enriched environments have been linked to neurogenesis.

It is important to note that these practices do not guarantee the prevention of all forms of cognitive decline associated with aging, the decline exists naturally, but they can contribute to overall brain health, and balance the scales by helping to maintain cognitive functions.

The Nutrition Variable

Without good nutrition the whole system is compromised, proper nutrition is critical for healthy brain development, especially during fetal development and early childhood. Essential nutrients like folate, omega-3 fatty acids, and various vitamins and minerals support brain growth and function. Malnutrition during critical periods can have long-lasting effects on cognitive development and brain health.

Proper brain function depends on a constant supply of oxygen and nutrients delivered through the bloodstream. Factors that affect blood flow and oxygenation, such as cardiovascular health and blood vessel integrity, have direct impacts on brain function. Conditions such as stroke or chronic hypertension can disrupt blood flow and lead to brain damage.

The brain is highly sensitive to inflammation and immune responses. Immunological factors can influence brain development and function, especially during periods of infection or autoimmune conditions. Neuroinflammation has been linked to several neurological diseases, including multiple sclerosis and Alzheimer's disease.

Aging, the aging process itself, is a physiological factor that affects the brain. As people age, natural changes in brain structure and function occur. These changes can include a reduction in brain volume, changes in neurotransmitter levels, and declines in cognitive abilities. Understanding the physiological aspects of aging is essential for dealing with age-related neurological conditions.

In summary, physiological factors are essential variables in the process of human development.

However, we are not done looking at everything, there are also ecological factors, which are environmental elements that have a significant impact on brain development, functioning and overall health. These factors encompass various aspects of the physical and social environment in which individuals live. Here are some of the key ecological factors related to brain development, for example the prenatal and early postnatal environment is critical. Factors such as maternal nutrition, exposure to toxins, stress during pregnancy, and access to prenatal care can all influence the development of the developing brain. Proper nutrition and a safe and nurturing environment during childhood are also crucial for healthy brain development.

Exposure to environmental toxins, pollutants, and contaminants can have detrimental effects on the brain. Lead, mercury, pesticides, and air pollutants are examples of substances that can negatively affect brain development, especially in children. These toxins can disrupt neural development and lead to cognitive and behavioural problems.

We've already mentioned nutrition, but why don't we also talk about diet: The availability and quality of food in the environment significantly affect brain development. Malnutrition, including undernutrition and overfeeding (obesity), can have adverse effects on cognitive development and brain function. Access to a balanced diet rich in essential nutrients is critical for optimal brain health.

Later, educational opportunities and the quality of education are critical ecological factors that influence brain development. Early childhood education, access to schools, and the availability of learning resources all play important roles in cognitive development. A stimulating educational environment can enhance cognitive skills and knowledge.

And accompanying this earlier development, socioeconomic status (SES) is a complex ecological factor that encompasses income, education, and occupation. A lower SES is often associated with limited access to resources and opportunities, which can affect brain development and lead to disparities in cognitive abilities and health outcomes.

Strong social networks and supportive relationships are essential for emotional and psychological well-being, which in turn can influence brain health. Positive social interactions can reduce stress, promote resilience, and enhance mental health.

On a day-to-day basis, life is accompanied by stress and adversity, exposure to chronic stress and adverse life events can have profound effects on the brain. Elevated levels of stress can lead to changes in brain structure and function, including overactivity of the stress response system. Childhood adversity, such as abuse or neglect, can have long-lasting effects on brain development and mental health.

To balance this, it is necessary to have at our fingertips the availability and quality of health services, they are critical ecological factors. Access to regular medical check-ups, early intervention for developmental disorders, and mental health support can have a significant impact on brain health and well-being.

An environment that encourages physical activity and recreation can contribute to brain health. Exercise has been linked to enhanced cognitive function, improved neuroplasticity, and a lower risk of neurodegenerative diseases.

Access to green spaces and natural environments can have positive effects on mental health and well-being. Spending time in nature has been linked to reduced stress, improved mood, and enhanced cognitive function.

Cultural norms, values, and community support systems play a role in shaping behaviours and attitudes related to mental health and brain development. Cultural practices, including traditions and rituals, can influence social and emotional development.

Consider the broader ecological context they are often necessary to support brain health and optimal cognitive function.

References chapter 4:

1. Knoll, A. H. (2014).
2. Life on a Young Planet: The First Three Billion Years of Evolution on Earth. Princeton University Press.
3. Passingham, R. E. (2016).
4. "How Good Is the Macaque Monkey Model of the Human Brain?" Current Opinion in Neurobiology, 40, 53-57.
5. Paus, T. (2010).
6. "Brain Development: The Changing Brain—Insights from Neuroimaging." In: Developmental Cognitive Neuroscience (pp. 7-20). MIT Press.
7. Barkovich, A. J., & Dobyns, W. B. (2018).
8. "Developmental neuropathology and the genetics of malformations of cortical development." Annals of the New York Academy of Sciences, 1426(1), 7-14.
9. Casey, B. J., & Jones, R. M. (2010).
10. "Neurobiology of the adolescent brain and behavior: implications for substance use disorders." Journal of the American Academy of Child & Adolescent Psychiatry, 49(12), 1189-1201.
11. Deoni, S. C. L., & Dean III, D. C. (2011).
12. "Remapping the cognitive and neural profiles of children who stutter: Commentary on van Lieshout and Peters (2010)." Journal of Fluency Disorders, 36(3), 220-223.
13. Johnson, M. H. (2011).
14. "Cerebral Cortex." Developmental Cognitive Neuroscience, 31-51.
15. Sowell, E. R., Peterson, B. S., Thompson, P. M., Welcome, S. E., Henkenius, A. L., & Toga, A. W. (2003).
16. "Mapping cortical change across the human life span." Nature Neuroscience, 6(3), 309-315.
17. Riggins, T., & Blankenship, S. L. (2011).
18. "Developmental Differences in fMRI Memory Paradigms: A Review." Child Development Perspectives, 5(3), 182-189.
19. Nelson, C. A., & Gabard-Durnam, L. (2020).
20. "Early Adversity and Critical Periods: Neurodevelopmental Consequences of Violating the Expectable Environment." Trends in Neurosciences, 43(3), 133-143.
21. Hackman, D. A., & Farah, M. J. (2009).
22. "Socioeconomic status and the developing brain." Trends in Cognitive Sciences, 13(2), 65-73.
23. Chaddock, L., Erickson, K. I., Prakash, R. S., Kim, J. S., Voss, M. W., VanPatter, M., ... & Kramer, A. F. (2010).
24. "A neuroimaging investigation of the association between aerobic fitness, hippocampal volume, and memory performance in preadolescent children." Brain Research, 1358, 172-183.
25. Bratman, G. N., Hamilton, J. P., Hahn, K. S., Daily, G. C., & Gross, J. J. (2015).
26. "Nature experience reduces rumination and subgenual prefrontal cortex activation." Proceedings of the National Academy of Sciences, 112(28), 8567-8572.
27. Kirmayer, L. J., & Swartz, L. (2011).

28. *"Cultural psychiatry in historical perspective." In Cultural Psychiatry: Euro-International Perspectives (pp. 1-52). Springer.*

29. *Kempermann, G., Frisén, J., & Gage, F. H.*

30. *"Neurogenesis in the Adult Brain I: Neurobiology."*

31. *Kempermann, G., Frisén, J., & Gage, F. H.*

32. *"Neurogenesis in the Adult Brain II: Clinical Implications."*

33. *Jennings, T. R. (n.d.).*

34. *The Aging Brain: Proven Steps to Prevent Dementia and Sharpen Your Mind.*

35. *Bottai, D. (n.d.).*

36. *The Brain-Derived Neurotrophic Factor (BDNF).*

37. *Khachigian, L. M., & Barrett, G. L. (n.d.).*

38. *Neurotrophic Factors.*

39. *Gibb, R., & Kolb, B. (n.d.).*

40. *The Neurobiology of Brain and Behavioral Development.*

41. *Ratey, J. J. (n.d.).*

42. *Spark: The Revolutionary New Science of Exercise and the Brain.*

43. *Medina, J. (n.d.).*

44. *Brain Rules: 12 Principles for Surviving and Thriving at Work, Home, and School.*

45. *Caine, J. B. (n.d.).*

46. *Aerobics of the Mind.*

Chapter 5

Family and development.

"Dutiful moms, doting dads.   Look at how the waters suddenly get dirty" **Gilberto Gil**

Grew up.

Today, we have all adapted to the pace of the era we live in and all that living in a world of limitless complexity entails. So far, we've discussed the cosmology of variables, the biology, and even the geology of variables. We talk about the complexity of each individual and look at embryonic development. Now it's the family's turn.

The term family itself is already synonymous with variations, here for mentioning some types of family constitution, however they are not a rigid model, families vary and evolve and adapt to anthropological and environmental issues.

So, we have:

Nuclear Family: is the most common type of family in many Western societies, consisting of parents and their children. It is characterized by the joint residence of parents and children.

Extended or Extended Family: includes relatives other than the nuclear parents and children, such as grandparents, uncles, aunts, cousins, etc. It can involve multiple generations living together or near each other.

Consanguineous Family: Based on descent and blood ties, this family structure emphasizes biological relationships. Family ties are key, and family ties are determined by bloodline.

The family, affinity, or marital structure is concentrated in the relationships created by marriage. The bonds between husband and wife are central, and the family is built around this marital unit.

Matriarchal and Patriarchal Family: Women are important in leadership and decision-making in matriarchal societies. In patriarchal societies, individuals occupy a central position in the family structure.

Compound Family: This is when people from different family groups come together to form a new family. This can occur through previous marriages, adoption, or other forms of family composition.

A single-parent family comprises only one adult who takes care of the children. This can be due to divorce, widowhood, or individual parenting choices.

Homoparental Family: It consists of same-sex parents and their children. It reflects the changing social and legal norms regarding parenting.

With the rise of globalization, many families are made up of members of distinct cultures and ethnic backgrounds.

Once again, this is just one example of the variability in the family, there are undoubtedly other models of lesser statistical significance, but I have mentioned the most expressive ones because it is a theme that I want to highlight.

The nuclear family is the basis of social development, where we are influenced by a myriad of variables. Each family is a cultural and socio-cultural microcosm; Each has its own perspective. Each family has its own unique "Constitution," shaped by its individual visions. This includes genetics, i.e. hereditary factors, the transmission of traits from one generation to another through genes. This influences not only physical traits, such as eye colour or height, but also genetic predispositions to certain medical conditions or behavioural traits and structures for cognitive development.

Here the variables occur in the evolution of mental abilities over time. It includes the acquisition and enhancement of skills such as memory and sensory perceptions, linked to memories (storage and retrieval of information), attention (focus and concentration), language (comprehension and expression), reasoning (information processing), and problem solving. In this ongoing process, the ability to manage and control one's emotions in response to different situations will appear. Everyone will learn to regulate their emotional life, manage, and control their own emotions in response to different situations. And there will be several ways in which emotions are communicated to others through gestures, facial expressions, body language, among other verbal or non-verbal behaviours. With this will come the recognition, understanding and management of one's own emotions and those of others, which is key to the consolidation of healthy interpersonal relationships.

The acquisition and values caused by the internalization of ethical norms and moral values influence ethical decisions and behaviours over time. Personality is formed from these characteristics already mentioned. It is within this development in the core of the family that consistent and enduring attributes influence a person's behaviour over time and in different situations. There are also several models created to detect variability in the formation of various personality traits, the main dimensions of personality include extraversion, neuroticism, openness to experience, agreeableness, and conscientiousness, according to the Big Five Factors model. It is at this point that a person sees themselves in terms of skills, characteristics, and personal identity.

Let's stop here and reflect on the role of variables, we see how they influence each day, each stage of life and each particularity of what makes each person up. Intra-family development and extra-family relationships develop, and more sophisticated forms of interaction begin to emerge that will also take on a very important dimension. Socialization shapes the formation of a person's identity by influencing socially accepted values, beliefs, and behaviours.

The appearance of impulses and goals can begin quite early in development when the social contrast arrives. They are the internal forces that drive behaviour, such as the pursuit of fulfilment, satisfaction, recognition, or specific goals. Learning or acquisition process encompasses the acquisition of knowledge, skills, and behaviours through experience, whether by observation, direct instruction, or interaction with the environment, where stress and adaptation will be important tools to form strategies and mechanisms used to deal with difficult situations and adaptation such as behaviour and cognition adjustments to face challenges and changes throughout life.

My reflection here is, what is the variability expressed in the family? How many individual differences caused by the influence of the family environment influence human nature in the various social sectors?

How does all this make up the psychological life and its performance in the environment in which it lives?

References chapter 5:

1. Costa, P. T., & McCrae, R. R. (1992). NEO PI-R Professional Manual. Psychological Assessment Resources.

2. Bronfenbrenner, U. (1979). The ecology of human development: Experiments by nature and design. Harvard University Press.

3. Erikson, E. H. (1959). Identity and the life cycle. International Universities Press.

4. McCrae, R. R., & Costa, P. T. (1999). A five-factor theory of personality. In L. A. Pervin & O. P. John (Eds.), Handbook of personality: Theory and research (Vol. 2, pp. 139–153). Guilford Press.

5. Baumeister, R. F., & Leary, M. R. (1995). The need to belong: Desire for interpersonal attachments as a fundamental human motivation. Psychological Bulletin, 117(3), 497–529.

6. Bowlby, J. (1982). Attachment and loss: Retrospect and prospect. American Journal of Orthopsychiatry, 52(4), 664–678.

7. Vygotsky, L. S. (1978). Mind in society: The development of higher psychological processes. Harvard University Press.

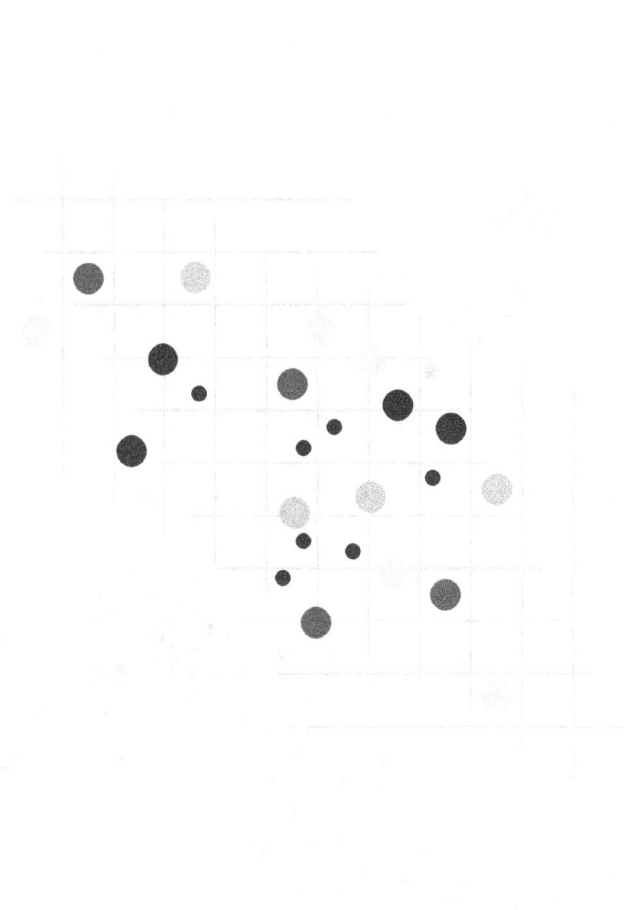

Chapter 6

Variables go to school.

"I don't teach my students; I only provide them with the conditions in which they can learn." **Seneca**

Learning

Here my purpose in talking about variables becomes more practical, I believe that the reader has already understood what I am getting at. Therefore, I want to share my dilemma, what makes me look at the world and often be unhappy with the lack of attention we give to the importance of understanding that no one is the same. It is in this inequality that development, the process, in which the construction of each one of us is made. And it is right at this time that we humans, still at a young age, will find the world outside leaving the family nucleus to confront social reality.

All those children, recipients of innocence and magnificent brains vacuuming information, ready to learn what happens outside the home with life.

Keeping the proper proportions and considering the socio-cultural contexts, the first contact with this new reality takes place in the learning centres, that is, in the schools.

It is of paramount importance to understand and evaluate the extent to which these nuclei are prepared or take into account the various variables already mentioned. Fortunately, it's not that challenging to identify a diversification-friendly mindset. However, resistance and an excessive inclination towards simplicity in education persist. This resistance does not necessarily stem from school centres, but rather from the incompetent management of governments that adopt simplistic policies, not contemplating education in a comprehensive way. Respect for individuality and the promotion of diversity should be central investment objectives.

Thus, we face a resistance to diversification in education often linked to simplistic policies and inadequate budgets, based on ignorance and manipulation of resources with unclear objectives. This is a global problem and is not just limited to low-resource countries.

But taking into account the awareness of the variables, we can dream of an ideal world where this awareness extends to the core of education, starting with teacher training, the goal is always, and on an ongoing basis, that they can recognize and meet the diverse needs of students, taking into account varied learning styles and promote inclusion.

It will be necessary to develop curricula that address topics related to cultural, ethnic, gender, and socioeconomic diversity. This makes children feel identified with the freedom to be unique and at the same time different enough to understand at a more advanced stage the importance of their social integration.

Teaching should avoid single, centralised approaches that do not consider the diversity of contexts. Developing assessment systems that consider multiple aspects of student achievement, going beyond standardized tests, thereby recognizing diverse skills and talents, going beyond traditional measures of academic success. Also including diverse perspectives across disciplines to reflect the breadth of human experience and its variations.

Promote school centres and environments that welcome diversity, without imposing conditions, promoting mutual respect, and understanding among students. Implementing anti-discrimination and anti-bullying policies to ensure that all students feel safe.

The participation of educators and students in the formation of the community and in educational decision-making, promoting a collaborative approach is the guarantee that in the future awareness will continue and be part of the surrounding social mindset.

Actively participate in dialogues about the importance of an education, and that it respects diversity, establishing partnerships with local organizations to support educational initiatives that meet the specific needs of the community. This is a strategy to ensure respect for individuality.

Ensure that schools are provided with adequate resources to meet the varying needs of students, including investments in physical facilities, learning materials, technology, and support for students with special needs. That should be the primary goal of these school communities. This is a social right and not a budget-dependent concession. From this point comes the promotion and awareness of society in defence of educational policies that encourage innovation and experimentation, allowing schools to adapt their teaching methods according to local needs.

References chapter 6:

1. Forbes, F. (Year of publication). Inclusive Education: Supporting Diversity in the Classroom.
2. Hammond, Z. (Year of publication). Culturally Responsive Teaching and The Brain.
3. Salend, S. J. (Year of publication). Creating Inclusive Classrooms: Effective, Differentiated and Reflective Practices.
4. Murphy, F., & Rubin, A. (Year of publication). Community Engagement, Organization, and Development for Public Health Practice.
5. Sharp, W. L., & Walter, J. K. (Year of publication). The Principal as School Manager.
6. Normore, A. H., & Lahera, A. I. (Year of publication). Innovations in Educational Change: Cultivating Positive Learning Environments.
7. Black, P., & Wiliam, D. (Year of publication). Assessment for Learning: Putting it into Practice.
8. Marshall, C., & Oliva, M. (Year of publication). Leadership for Social Justice: Making Revolutions in Education.
9. Freire, P. (Year of publication). Pedagogy of the Oppressed. Originally published in 1968.

Chapter 7

The variables will work.

"Do your job masterfully. It will then become a permanent part of your life."
Ralph Wando Emerson

Let's get to work.

When reflecting on the wide range of possibilities and the complexity inherent to each human being, one of the most challenging moments in existence arises: the need to adapt to the resources of modern life. This challenge materializes when it is time to move from the study environment to the work environment, often facing the demand of performing both activities simultaneously, depending on the socioeconomic context and the geographical situation in which everyone finds himself. This dynamic requires a considerable effort, representing a real battle to position life in such a way as to continue progressing and, therefore, contribute to the continuity of society.

In the corporate scenario or in the world of work, especially in societies rooted in capitalism, the priority falls on the enrichment of companies and companies. These entities are conceived for the purpose of generating resources by continuously increasing their earnings in favour of their owners. While some businesses can benefit the community or region, either by creating jobs or producing goods that make human life easier, the core of these organizations is directed toward steady growth and maximizing benefits. It is important to highlight that companies do not wait to receive uniform individuals; rather, they are faced with the flow of a multitude of personal variables, each person unique in his or her own diversity.

In this context, companies with no idea what to do, copy old models passed down through generations, often offering a meld into which people are expected to fit regardless of their diverse ethnicities, perspectives, cultures, and ways of thinking. Although in some recent context's companies have shown concern about these issues, it is crucial to recognize that, in a large part of the planet, a work model prevails that disregards the differences that characterize us as human beings.

From training centres to the nuclear family, companies should welcome individuals who join their ranks, not only to produce innovative goods and technologies, or even to take care exclusively of their benefits, but also to be part of a social whole to promote a work environment that respects and values diversity, thus contributing to a fairer and more inclusive society. However, the great reality is that in the workplace, we are faced with a considerable cultural clash between each individual and the very essence of the company.

The confrontation manifests itself in the company's expectations, in the method used to adapt and classify employees, in the leaders involved and, in the training, aimed at conducting the recruitment and integration process. This clash contrasts with the intrinsic diversity of each human being who enters this scenario, preparing to conquer a position within a stylized and predefined culture that, in the first instance, aims at its own successes and enrichments.

When entering the work environment, most workers are faced with negative experiences, especially when inserted in unprepared environments. In these situations, it is common to find the absence of ethics and integration, revealing a clear separation between the company and the employees. The dynamic is often established as "we are the companies, and you are the employees", a position that generates discontent and contempt for individualities, because people do not feel integrated, they do not even feel inside, the feeling is confused, they are inside the company, but they are not the company, they are and they are not, or any variation of the same model fits.

In many cases, the company makes evident its lack of honesty in relation to its intentions with its employees, reinforcing the dichotomy between "us" and "you", thus maintaining an eternal servant culture, where "you" are just fuel for a machine that will not change.

The lack of impartiality is notable, as the company usually defines itself as a separate entity, making it clear that there is a privileged group, integrated only by the fact that it belongs to the same organization. This attitude contributes to workers feeling marginalized, amplifying the feeling of lack of integration.

Thus, there is a constant tension between the individual identity within the company and the need to produce according to the expectations of the organization. The representation of the individual is intrinsically linked to their ability to meet the demands of the company. However, understanding what exactly constitutes this production becomes challenging, since the concept is constantly changing, influenced by the variability of each customer served by the company.

This variability in production is compounded by the lack of an organizational culture that promotes adaptation. The lack of a humanistic approach, concerned with the well-being of each person and their professional development, contributes to the difficulty of adaptation and the feeling of lack of support for individual growth within the company.

However, all is not lost. Increasingly, it is understood that the core of a company is made up of people, and these people are complex sets of variables. As business owners recognize this reality, they have a powerful tool in their hands. The realization of this knowledge reveals itself as the key to innovation, the discovery of a blue ocean, and the achievement of comprehensive success.

To this end, it is imperative to conduct personalized assessments to understand the needs, work styles, and development goals of each team member. Considering differences in cognitive, emotional, and motivational styles is crucial for adapting expectations and offering individualized support.

Giving importance to skills, it is essential to assess individual skills, identifying strengths and areas for development. The study of specific and contextualized tools, such as competency maps, helps in visualizing the skill set of employees.

A powerful tool to promote health at work and eliminate anxieties and constant feelings of panic is the identification of each member's cognitive styles. By recognizing visual, auditory, or tactile learning styles, it is possible to adjust the delivery of information and tasks to optimize comprehension and retention. Lack of adequate preparation is often the source of insecurity regarding personal performance. It's crucial to be mindful of employees' emotional needs, recognizing factors such as resilience, communication preferences, and stress management. Emotional support proves to be just as important as a sales strategy and should be integrated into the company culture. Equally important is to adjust the leadership approach according to the individual emotional characteristics of the company's sectors.

Based on this understanding, the company should strive to collaborate with each member, setting clear and individual goals that are aligned with organizational objectives. By setting realistic and challenging goals, cohesion is fostered, considering everyone's specific aspirations and abilities.

Fostering a work environment that celebrates diversity and promotes inclusion, recognizing and valuing the individual differences of team members, is evidence that the company is a living, thinking organism, not a simple machine. Offering training programs that consider diverse learning styles and personalizing professional development plans based on everyone's specific educational trajectories and career goals is key to promoting career longevity and engagement.

Adopting communication strategies that consider variability in communication styles ensures clarity and understanding. This practice encourages an open environment where different perspectives can be expressed and valued.

Respecting human variability in a company, it is crucial to develop leaders who possess adaptable leadership skills, able to recognize and embody diverse leadership styles, including inclusive leadership, which values the unique contribution of each team member.

Implementing policies that allow for flexibility in the work environment, recognizing the different needs of work-life balance, is essential. Creating environments that accommodate different work styles, providing options for collaboration and independent work, contributes to building a healthier and more productive environment.

Individually tailored mentoring programs are a strong demonstration that the company values sharing experiences and growth, treating people with direct communication, and considering everyone's professional development goals. Incorporating cross-mentoring, where professionals from different areas can share experiences and insights, is a powerful tool for respect and integration.

Perhaps these are some ideas, certainly already elaborated by others who are concerned with the impact and importance of variables and what they mean for individuality.

References chapter 7:

1. Cox, T. (1994). Cultural diversity in organizations: Theory, research, and practice.

2. Thomas, D. A. (1991). The impact of cultural diversity on organizations.

3. Cox, T., & Blake, S. (1991). Managing cultural diversity: Implications for organizational competitiveness.

4. Schein, E. H. (1992). Organizational culture and leadership.

5. Bass, B. M., & Riggio, R. E. (2006). Transformational leadership.

6. Goffee, R., & Jones, G. (1996). What holds the modern company together?

7. Buckingham, M., & Clifton, D. O. (2001). Now, discover your strengths.

8. Tannenbaum, S. I., & Yukl, G. (1992). Training and development in work organizations.

9. Goldsmith, M., & Reiter, M. (2010). What got you here won't get you there.

10. Greenhaus, J. H., & Beutell, N. J. (1985). Sources of conflict between work and family roles.

11. Kossek, E. E., & Lambert, S. J. (2005). Work and life integration: Organizational, cultural, and individual perspectives.

12. Friedman, S. D., & Greenhaus, J. H. (2000). Work and family—Allies or enemies? What happens when business professionals confront life choices.

13. Goleman, D. (1995). Emotional intelligence.

14. Kram, K. E. (1985). Mentoring at work: Developmental relationships in organizational life.

15. Ragins, B. R., & Kram, K. E. (2007). The Handbook of Mentoring at Work: Theory, Research, and Practice.

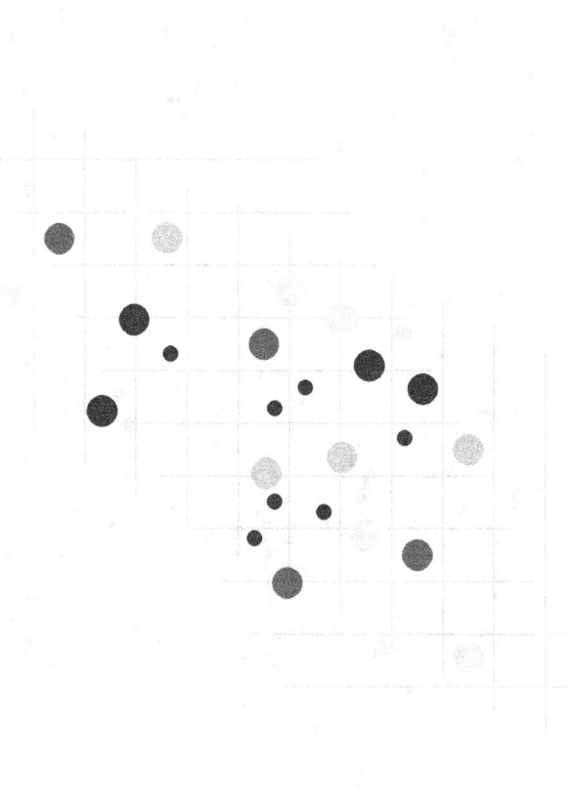

8 - Final chapter

Nothing is the same.

"Always remember that you are absolutely unique. Just like everyone else."

Margaret Mead

Diversity

We must remember that variables accumulate. With each passing minute, time carries countless variables that accumulate and sediment in our past. Some disappear, others endure, but every minute, everything we mentioned above is proof that we live surrounded by variables.

From the formation of our planet, the development of geological and biological ages, through the appearance of life and the development of this life, to the present day, from the macro to the micro, no one is the same. No one has ever been the same, and no one will ever be the same, because the variables are infinite.

We tend to put everything in the same form, classify things so that we can understand, give names to situations, behaviours, and attitudes, always thinking about categorization. However, all this categorization is futile.

Life, at its core, is governed by fundamental principles of chaos and entropy. Just as in physics, where the second law of thermodynamics reveals that the entropy of a closed system tends to increase over time, in human existence, we witness constant unpredictability and complexity.

Systems, whether physical or social, are dynamic and susceptible to change. The slightest initial change can trigger significantly different outcomes, giving rise to the ripple effect of chaos. This notion of systems in constant motion and subject to continual change reflects the interconnectedness of disorder and evolution.

As we consider the application of these principles to human life, we realize that we are the product of this constant movement. Each person, regardless of whether they are labelled as good or bad, friend or foe, represents a manifestation of the complexity inherent in social systems. Such labels are, in fact, social constructions or models created in our feelings, underlining the subjectivity of these categories.

The difficulty in accurately predicting the future, evidenced by the idea that small changes can result in unpredictable consequences, highlights the inherent limitations of predictability and determinism. Life, as a closed system, is prone to move toward more disordered states, defying inflexibility to predictability.

By embracing complexity as a wonder, a deep admiration for each person emerges, recognizing everyone's uniqueness and unique contribution to the vast human landscape. This acceptance of diversity and the understanding that order can evolve into chaos and vice versa are philosophical principles that echo through the ages, reminding us that change is a fundamental constant in existence. Ultimately, we are an integral part of dynamical systems, where evolution is the only certainty.

A challenge. On the ladder of importance, the highest rung is ours, it is up to us to have a passion for science. As I have already mentioned, think critically, get to work, study, and participate in everything that promotes and makes our variables viable, so that they can be used to their full potential.

Here I leave my philosophical daring, in the end, the book is mine.

The human being is nothing more than a set of variables, forged in the cosmos, expressed on earth, assembled in the womb, and known in the mind.

Enjoy life and know that you will change a lot.

We're all going to change.

References chapter 8:

1. *Pollard, T. D., & Earnshaw, W. C. (2002). Cell biology. Saunders.*
2. *Alberts, B., Johnson, A., Lewis, J., Raff, M., Roberts, K., & Walter, P. (2002). Molecular biology of the cell. Garland Science.*
3. *Kandel, E. R., Schwartz, J. H., & Jessell, T. M. (2000). Principles of neural science. McGraw-Hill.*
4. *Ridley, M. (2004). Evolution. Blackwell Publishing.*
5. *Bouchard, T. J., Jr., & McGue, M. (2003). Genetic and environmental influences on human psychological differences. Journal of Neurobiology, 54(1), 4-45.*
6. *Plomin, R., DeFries, J. C., Knopik, V. S., & Neiderhiser, J. M. (2016). Top 10 replicated findings from behavioral genetics. Perspectives on Psychological Science, 11(1), 3-23.*
7. *Turkheimer, E. (2000). Three laws of behavior genetics and what they mean. Current Directions in Psychological Science, 9(5), 160-164.*
8. *Prigogine, I., & Stengers, I. (1984). Order out of chaos: Man's new dialogue with nature. Bantam.*
9. *Gleick, J. (1987). Chaos: Making a New Science. Viking.*
10. *Kauffman, S. A. (1995). At Home in the Universe: The Search for Laws of Self-Organization and Complexity. Oxford University Press.*
11. *Hilborn, R. C. (1994). Chaos and Nonlinear Dynamics: An Introduction for Scientists and Engineers.*
12. *Kuhn, T. S. (1962). The Structure of Scientific Revolutions.*
13. *Mitchell, M. (2009). Complexity: A Guided Tour.*
14. *Hofstadter, D. R. (1979). Gödel, Escher, Bach: An Eternal Golden Braid.*